WHEELS
AT WORK AND PLAY

ALL ABOUT

TRACTORS

Wheels at Work and Play
All about Diggers
All about Motorcycles
All about Race Cars
All about Special Engines
All about Tractors
All about Trucks

Library of Congress Cataloging-in-Publication Data

Stickland, Paul.
All about tractors / Paul Stickland.
p. cm. — (Wheels at work and play)
Summary: Depicts different kinds of tractors and their various uses.
ISBN 0-8368-0423-6
1. Tractors—Juvenile literature. [1. Tractors.] I. Title. II. Series.
TL233.S695 1990
629.225—dc20 90-9817

This North American edition first published in 1990 by
Gareth Stevens Children's Books
1555 North RiverCenter Drive, Suite 201
Milwaukee, Wisconsin 53212, USA

Series editor: Tom Barnett
Designer: Laurie Shock

Printed in the United States of America

1 2 3 4 5 6 7 8 9 96 95 94 93 92 91 90

WHEELS
AT WORK AND PLAY

ALL ABOUT
TRACTORS

Paul Stickland

Gareth Stevens Children's Books
MILWAUKEE

The tractor plows fields.
They are ready for planting.

The sea gulls look for worms.

This huge tractor pushes and pulls.

Tractors break down.
They need to be fixed.

Mirrors help the driver see.

This man takes his sheep to market.

This trailer is full of straw.

It will make warm beds for the animals in the winter.

The green tractor has
a digger.

It loads dirt into the
red trailer.

Tractors come in many different shapes and sizes.

Glossary

digger
A tool that can be attached to a tractor that digs holes.

market
A place where people can buy and sell goods.

plow
To dig up the earth. Also, a tool, pulled by a tractor, that digs up the earth.

tractor
A type of truck used in farming.

trailer
A cart pulled by a tractor.

Index